The Way You L

FIRST EDITION

Sanna and Emese are experienced teachers and mothers who felt the need to bring both roles together when their children's mainstream learning required supplementing. See what they say about their first book "You Can Teach Me Maths!"

Foreword

When our little ones entered primary school, two things quickly became clear. One, the pace of the curriculum was too fast for kids who were still lost in their fantasy world. Two, that even the more abled kids lacked the opportunity to deepen their new skills well enough to be able to apply them further.

Maths skills build on each other; one wobbly block can make the whole structure tumble. Therefore, children need enough time building a solid foundation to avoid learning gaps. Being in the education sector, we knew there was no easy fix to filling in the gaps. There was not going to be a ready-to-buy solution in the form of a practice book or a custom-made app. We found that these often lack input and children cannot practice something they do not know. So, we rolled up our sleeves, carved out some time for our kids and this book is a collection of the maths sessions we designed for them.

You Can Teach Me Maths!

The minute your child was born, you became their first teacher. As an advantage, nobody knows them as well as you do; nobody can read them or adapt to their needs as well as you can. Therefore, we believe you can teach your child. However, you will need to ask yourself the following questions:

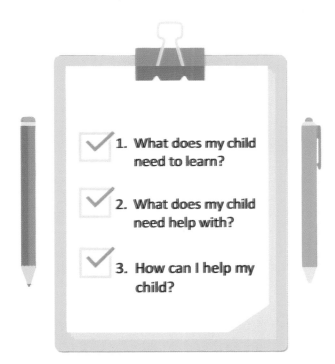

1. What does my child need to learn?

2. What does my child need help with?

3. How can I help my child?

You may not have the answer to any of these questions and that is fine. 1. Your child is expected to know skills outlined in the national curriculum. If you accept too little, your child will fall behind; if you demand too much, your child might struggle. 2. Work your way through this book and focus on the skills that they are still a bit shaky with; these may need to be revisited. 3. Once you have identified the skills that your child needs help with, it is likely that they need input, and you need to teach. This book will help with all of these.

How to use this book?

Each of our sessions is based around the context of a job. For example, your child will be a car surveyor, a chef, or even a birdwatcher. They will enjoy learning while playing make believe, but most importantly, they will remember the skills they need in order to play along. To make this happen, you need to follow a simple structure as explained below.

Here you introduce the context i.e.: jobs

Car Surveyors

Outcome: understand and create tally charts and simple tables
You need: a ruler, pencil and some paper

Set Up

Tell your child that today they get to be surveyors and find out what car colours are the most/least popular on their road.

Focus

At this stage your child has some fun with the skill.

Explain to your child that sometimes we use tallies to count. Ask your child to:
- have a look at the following tallies in Picture 1
- write down the number each tally represents
- notice what is different about the fifth tally
- think about why this is useful (answer: makes counting in 5s easier)
  Picture 1
- write tally marks for 6, 7, 8, 9 and 10.

For surveys we use tally charts. To make one:
- use a ruler and draw a neat table with 3 columns
- add as many rows as colours you want to work with
- create a simple table like in Picture 2.

Colours	Tally	Total
black	III	9
silver		
pink		
etc.		
Total		

Picture 2

This is where you help your child understand the skill they need to play along.

Activity

Let the surveying begin! Take your child up the road and complete your tally chart.

Follow up

Now your child knows how to create a tally chart. They also know how to read information from the table they just created. Ask them to report back on the most/least popular car colours on their road to a member of the family.

Here your child finds different ways to use the skill they just learned.

Why the cardboard box on the cover?

You might have noticed the cardboard box on the cover of this book. It comes from an idea which has been deeply rooted in our teaching practice, but also when working with our own children. A cardboard box is basic, and it can become anything you would like it to be. Similarly, a lesson where you do not give anything (i.e. worksheets), opens opportunities for learners to make that lesson what they want it to be. Therefore, our sessions require minimal resources, so that learners take ownership over their learning and are involved in creating their own materials. When you engage a learner's imagination, you will multiply the success of your lesson. Not convinced? Let us look at our Athletes Unit:

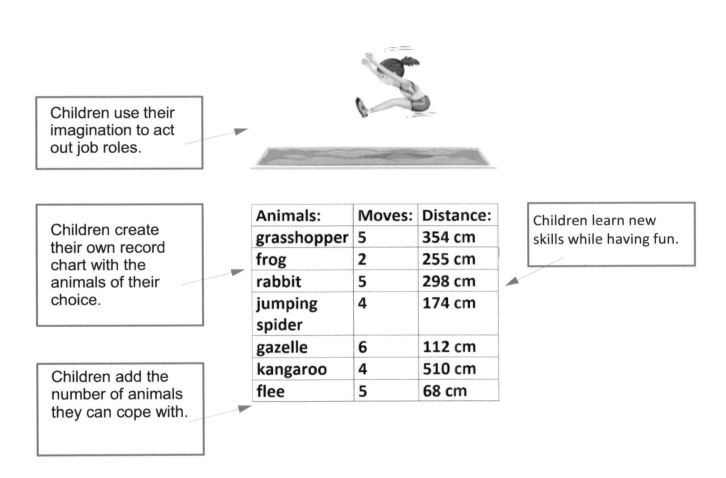

Children use their imagination to act out job roles.

Children create their own record chart with the animals of their choice.

Children add the number of animals they can cope with.

Children learn new skills while having fun.

Animals:	Moves:	Distance:
grasshopper	5	354 cm
frog	2	255 cm
rabbit	5	298 cm
jumping spider	4	174 cm
gazelle	6	112 cm
kangaroo	4	510 cm
flee	5	68 cm

The above examples show that each lesson is personalised, differentiated and gives ownership to your child. All these result in better engagement which leads to greater outcomes.

To make these lessons as fruitful as they can be look at the following tips:

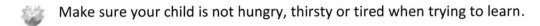

- 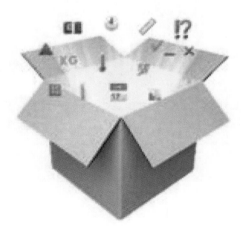 Make sure your child is not hungry, thirsty or tired when trying to learn.

- Offer breaks as and when needed.

- Reassure your child that it is OK to make mistakes, because it is a sign that they are learning.

- Be cautious when using sharp objects and be mindful of your surroundings when outdoors.

- Show love, patience and reassurance. Use these sessions as opportunities to bond with your child.

- Take your time and go at your child's pace.

- Praise your child on effort and welcome their suggestions to solving problems.

- Celebrate every little success with enthusiasm and positivity.

- Remember to have fun!

Making time

Your child's learning is vital, so you need to find time for it. Start by looking at your existing routine. When does your child learn best? When are you least likely to be distracted by other commitments? Once you have found the perfect time, share this new schedule with your child and form a visual plan together. You could try creating a map or a display of the new routine. This will promote ownership and your child will know what is coming and when. Once all set, all you need to do, is stick to it!

The Way You Learn

The Way You Learn Ltd

As our children entered primary school, we felt the need to support their mainstream learning. We started to design activities for our children that worked well for their needs. In 2019 we formed The Way You Learn Ltd. with a simple goal in mind: to help parents like us support their children outside of school.

We hope you will enjoy working through this book with your child. We are looking forward to reading your positive reviews. If you have any questions or comments, please email us at info@thewayyoulearn.co.uk

A portion of our profits will be used to reach out to children around the world who may not have access to education. Thank you for helping us achieve this.

Apology

Although all attempt was made to avoid any mistakes, we would like to apologise in advance if a few slipped through. We are sure that you, more than anyone, will understand how difficult it is to work at home with kids around. We put this book together outside of our working hours, between nappy changing, preparing meals and addressing the needs of crying or happy kids. Not to mention the time difference between the US and the UK, which resulted in one of us being starved for breakfast while the other one was late for dinner. Despite all these struggles, working on this book brought us great joy and satisfaction. It is now your turn to have some fun!

Contents

Users around the world

For those of you who may not be familiar with the units of measure we use in the UK, you will be happy to know that our sessions can be easily converted to match the units that are used where you are. Please see the suggestions below for some of our units.

Unit 10   **Patisserie Chefs**
Outcomes: measurement in pounds or ounces

Unit 11   **Athletes**
Outcomes: measurement in inches/feet/yards

Unit 12   **Weather Presenters**
Outcomes: measurement in Fahrenheit

Unit 13   **Mocktail Mixers**
Outcomes: measurement in pints or quarts

With that said, it is now time to teach maths!

Ancient Stock Takers

Outcome: place value of ones, tens and hundreds
You need: a small cardboard box, some wire, pasta or beads

Set up

Tell your child that today they will time travel back to ancient times and:
- ask if they know what people used to count with
- show them a picture of an abacus
- tell them that this can help with calculation and today they are going to make one.

To create your own abacus:
- find a small cardboard box and cut 3 small holes on both sides opposite each other
- thread a wire through, putting 10 beads or pasta on each wire
- make sure you secure each end and the wire is stretched.

Focus

Using the abacus:

- put a handful of raisins, dry beans or pasta in a bowl
- arrange all the beads on the abacus to the left
- start counting your items one by one moving the beads on the first wire to the right
- when you get to ten, shift all the ones back to the left
- now on the second wire move 1 bead to the right, this means we have counted
10 items so far (*see figure*)
- continue counting on the first wire, when you get to 20 repeat the above actions:
i.e.: push all the ones beads back to the left and 1 tens bead to the right on the second wire.

Activity

Let's stock take:
- create a table with as many rows as you want to work with
- count a variety of items in the kitchen
- choose some items that are less than 10 and some that are more than 10
N.B.: in KS1 pupils only learn numbers up to 100
- keep an accurate record of these (*see figure*).

name of items	number of items	number of tens	number of ones	number sentence
stock cubes	32	3	2	32=30+2
tomatoes	7	0	7	0+7=7
dry pasta	57	5	7	50+7=57
etc				

Follow up

Take your abacus around the house and count anything you like: pens, pencils, books, matchboxes, Lego blocks etc.

Card Game Makers

Outcome: use number bonds* of 10 in several forms
You need: paper, colouring pens or pencils, and scissors

Set up

Ask your child what card games they like to play. Tell them that they will be designing their own card game today.

Prepare the cards:
- cut 36 cards of the same size
- create 4 sets of cards numbered 1 through 9.

1	2	3	4	5	6	7	8	9

Focus

To begin:
- shuffle the 4 sets of cards together
- create as many number bonds of 10 as you can
- put these number bond cards as pairs

(possibilities include: 1 + 9, 2 + 8, 3 + 7, 4 + 6, 5 + 5, 6 + 4, 7 + 3, 8 + 2, 9 + 1).

Activity

Design your cards by colouring, drawing or decorating with stickers. Then play Snap:
- deal cards equally between you and your child
- turn both your cards face down
- take turns turning over your top card
- add the 2 numbers (one number from the adult's card and one number from the child's card)
- when the numbers add up to 10, shout "snap"
- the person who says "snap" first keeps the pair
- play until you cannot make any more number bonds of 10.

Follow up

The next step is to recognise the relationship between these number bonds. Help your child understand that the bonds also apply to subtraction. Some examples include: 10 – 6 = 4; 10 – 4 = 6. Then:
- make two additional cards: one that has "**10 –**" and one that has "**=**"
- place these between the 2 piles of number cards
- play the game using the same steps as above, but this time use subtraction facts.

10 -	6	=	4

*Number bonds are an important problem-solving strategy. Children need to know number bonds 1 to 10 by rote so once your child understands the concept and is familiar with the possibilities, they are ready for the next level. Try number bonds for different values. *Happy Bonding!*

Board Game Makers

Outcome: use multiplication facts (we use the 5 times table as an example)
You need: A3 and A4 paper, scissors, a 12-sided dice
(search online for a free printable template)

Set up

Your child needs to understand *why* we use times tables. To begin:
- give your child several raisins
- say to your child *suppose your friend wants 5 raisins*
- ask your child to count 5 raisins
- ask your child to draw a pictorial representation of 5 such as ::.
- ask your child to write this scenario down as 5 x 1 = 5.

Focus

Tell your child that two of their friends would like 5 raisins each. Now:
- ask how many raisins would be needed- your child counts out 5 raisins twice
- draw a pictorial representation of this scenario such as . . . . or : : : : :
- write this as an equation (5 + 5 = 10)
- write this as a multiplication equation (2 x 5 = 10 **or** 5 x 2 = 10)
- continue until your child fully understands the concept.

5 + 5 = 10

2 x 5 = 10

Activity

To prepare the game:
- write down your chosen times tables up to 12
- copy the answers on a piece of paper
- refer to these as number cards
- create your board game with minimum 20 spaces.

5	10	15	20	25	30
35	40	45	50	55	60

To play the game:
- place a counter of your choice on the start space
- move one space forward when it is your turn
- roll the 12-sided dice and multiply the number by 5
- find the number card with the answer and keep it
- play again
- try to collect all 12 number cards before you reach the finish.

Follow up

Keeping the card game in mind, think about asking your child these questions:
> What number cards were you **not** able to pick up, if any?
> What number should you have rolled to be able to pick those number cards up?

Y2 When ready, you can use it for other times tables like 2s and 10s.

Housekeepers

Outcome: use number bonds of 10 and 20 in several forms
You need: paper, colouring pencils, minimum 10 clothes pegs, and a clothes hanger. For extra fun you will need a bunch of small pegs and approximately 2 metres of string.

Tell your child that they are going to be housekeepers and today they need to do the laundry.

Focus

Y1 Target value 10:
- ask your child to put 10 clothes pegs onto your hanger
- explain that each peg has the value of 1
- ask them to show different representations of the target value using the pegs
e.g.: move 3 pegs on one side, 7 remain on the other
  - record your number bond variations on paper e.g.: 3+7=10
(other possibilities: 1 + 9, 2 + 8, 3 + 7, 4 + 6, 5 + 5, 6 + 4, 7 + 3, 8 + 2, 9 + 1)
- cut 18 different shaped cards in the form of different clothing as below
- write numbers 1 to 9 **twice**.

Y2 Target value 20:
- ask your child to put 20 clothes pegs onto your hanger
- explain that each peg has the value of 1
- ask them to show different representations of the target value using the pegs
e.g.: move 13 pegs on one side, 7 remain on the other
  - record your number bond variations on paper e.g.: 13+7=20
(other possibilities: 1 + 19, 2 + 18, 3 + 17, 4 + 16, 5 + 15, 6 + 14, 7 + 13, 8 +12, 9 + 11, 10+10)
- cut 20 different shaped cards in the form of different clothing as above
- write numbers 1 to 19
- add one more card of 10.

Activity

Activity Design your clothes by colouring, drawing or decorating with stickers. Then:

- mix all the clothing cards and put them in a box i.e.: your washing machine
- empty the washing machine placing your clothing on to a washing line
- pin up clothes together that add up to the target number bond
- when all complete, clothes can be folded into a basket together.

Follow up

Number bonds are an important problem-solving strategy. Children need to know number bonds 1-20 by rote so once your child understands the concept try number bonds for different values. *Happy Bonding!*

Birdwatchers

Outcomes:

- Y1 counting in 2s, 5s and 10s
- Y2 counting in 2s, 5s and 10s, from any given number on the 100 number line

You need: paper, colouring pencils

Set up

Tell your child that some people watch birds for hobbies, others for their work. Say that today you are going to be looking at birds:

- ask how many different kinds they can spot outside
- write a list e.g.: a sparrow, a pigeon, a robin, a crow etc.

Focus

Yr1 Create your record chart:

- draw a table (*see figure 1*) with the birds of your choice
- decide and write a score of either 2, 5 or 10 for each of your birds
- when your child spots a bird they write this score in the table
- continue scoring each time you spot a bird until fun
- when finished, your child should count in 2s, 5s or 10s to get their total scores.

Yr2 Create your record chart:

- draw a table (*see figure 2*) with the birds of your choice
- decide and write a score of either 2, 5 or 10 for each of your birds
- when your child spots a bird, they earn a score and start by adding it to 0
(e.g. 0+5= 5 +2= 7 + 10= 17)
- continue adding each time you spot a bird
- as numbers grow, encourage your child to put the bigger number in their head, and use their fingers for adding the smaller number.

Let the birdwatching begin! With your record chart in hand, take your child for a walk to the park or sit by the window and:

- start scoring the birds as you spot them
- if need be, adjust your list by adding more or fewer birds to the scoring chart.

You can create new charts for:

- cars, trucks, and buses
- flowers, trees and bushes
- dogs, cats and even people.

 sparrow 2 points	 pigeon 2 points	 magpie 5 points	 crow 5 points	 robin 10 points
2				
2	2			
Total:	Total:	Total:	Total:	Total:

figure 1

 sparrow 2 points	 pigeon 2 points	 magpie 5 points	 crow 5 points	 robin 10 points
example: 0+5= 5 +2= 7 + 10= 17				
				Total:

figure 2

Gardeners

Outcome: use multiplication and division facts for the 2s, 5s and 10s
You need: seeds (or substitute), scissors, colour pens and paper

Set up

Y1 Tell your child that they will be gardeners today. First:
- draw **two** flower heads
- give your child 20 sunflower seeds
- ask them to drop one seed at a time into the circles until all the seeds are gone
- count the flower seeds and check that each flower head has the same number of seeds.

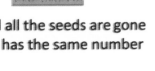

Y2 Tell your child that they will be gardeners today. First:
- draw **five** flower heads
- give your child 20 sunflower seeds
- ask them to drop one seed at a time into the circles until all the seeds are gone
- count the flower seeds and check that each flower head has the same number of seeds.

Focus

Ask your child a few questions:

Have you divided the seeds equally amongst the flower heads?
How many seeds do you have on each flower head?
Can you write these out with numbers and symbols?

Y1 answers: 10 + 10 = 20   20 = 10 + 10
Y2 answers: 20 ÷ 5 = 4   5 x 4 = 20

Activity

Let's play "naughty bird!"
Decide on the number of flower heads you want to work with e.g.: 2, 5 or 10 and draw that number of flower heads. You will also need to draw a scarecrow.

Tell your child they will be gardeners. Today they will be planting seeds, but need to watch out for the naughty bird who might be picking and dropping off seeds. Whilst planting they can put up their scarecrow to keep the naughty bird away. To start the game, ask your child to collect a handful of seeds.

Activity

Gardeners (child):
- put up your scarecrow and count the seeds that you have
- divide these equally amongst the flower heads
- when you have left over seeds, just put them aside
- look at your flowers heads and write these out with numbers and symbols (*see focus*)
- to keep score, for every correct answer draw a reward flower on the side
- see how many reward flowers you can collect in 5 minutes
- when you have finished, take down your scarecrow.

Naughty Birds (parent):
- watch the gardener plant seeds equally on the flower heads
- wait until they have taken down their scarecrow
- now, you can take away or drop off more seeds for your gardener to work with.

Ready, steady, plant!

Follow up

An alternative division technique is using a number line. Why not try the "molehill" method?
To find the answer for 20 ÷ 5 draw a number line to 20 and then:
- put your pencil on number 20
- count 5 spaces back and draw a big molehill
- continue in this manner going back to 0
- count the number of hills (4)
- this is your answer!

Pizzaiolos-pizza makers

Outcomes:
- **Y1** simple fractions (1/2 and 1/4) of objects
- **Y2** fractions (1/3, 1/4, 2/4 and 3/4) of objects and quantities

You need: paper, colour paper, pencils and glue or playdough   Part 1

Set up

Ask your child about their favourite pizza, then:
- ask about possible toppings and make a list of these
- create the toppings from paper, playdough or any other alternatives.

Focus

Y1 Cut out 2 large paper circles as your pizza bases and:
- tell your child to share it with a friend
- ask them to divide one base into 2 pieces with a pencil
- establish that 1 pizza cut into 2 pieces is ½ = 2 halves (*see figure 1*)

- take a new "pizza" base and ask your child to share the pizza with 3 friends
- ask them to divide one base into 4 pieces with a pencil
- look at the pieces together and establish that 1 pizza cut into 4 pieces is ¼ = four quarters (*see figure 2*).

Y2 Cut out 3 large paper circles as your pizza base and create a menu:
- ask your child to create a 12-topping pizza for **2** (*see figure 3*)
- child takes a pizza base and divides it into 2 halves with a pencil
- child chooses 2 types of toppings and works out that each side should have **6** toppings each
- ask them to count the toppings on each slice, **establish that ½ of 12 =6**
- give this creation a name and write it on the menu

e.g.: Pepperoni Delight for 2 – toppings: ½ tomatoes (6), ½ pepperoni (6)

- ask your child to create a 12-topping pizza for **3** (*see figure 4*)
- child takes a pizza base and divides it into 3 sections
- child chooses 3 types of toppings and works out that each slice should have **4** toppings each
- ask them to count the toppings on each slice, **establish that 1/3 of 12 =4**
- give this creation a name and write it on the menu

e.g.: Fishy Heaven for 3 – toppings: 1/3 sardines (4), 1/3 olives (4), 1/3 tomatoes (4)

| *figure 1* | *figure 2* | *figure 3* | *figure 4* |

- ask your child to create a 12-topping pizza for **4** (*see figure 5*)
- child takes a pizza base and divides it into 4 sections
- child chooses 4 types of toppings and works out that each side should have **3** toppings each
- ask them to count the toppings on each slice **establish the fact that ¼ of 12 =3**
- give this creation a name and write it on the menu

e.g.: Feast Lovers for 4- toppings: ¼ sardines (3), ¼ olives (3), ¼ tomatoes (3), ¼ pepperoni (3).

figure 5

Activity

Children are now ready to play pizzeria. Using their menu, they should take orders, create their pizzas and deliver them to their customers. *Buon appetito!*

Follow up

Make sure your child grasps all of the above before moving on to Part 2: fractions of quantities. If need be, leave it for another day.

Pizzaiolos-pizza makers

Outcomes:
- **Y1** simple fractions (1/2 and 1/4) of quantities
- **Y2** fractions (1/3, 1/4, 2/4 and 3/4) of quantities

You need: paper, coloured paper, pencils and glue or playdough   Part 2

Set Up

Ask your child to recall what they did in the last pizza activity. Together create the toppings again or find the ones you made last time.

Focus

Y1 Create a pizza menu:

12-topping pizza for **2** (*see figure 6*)
- child takes a paper pizza base and divides it into 2 halves
- child chooses 2 types of toppings and works out that each slice should have **6** toppings each
- ask them to count the toppings on each slice, **establish that 1/2 of 12 = 6**
- give this creation a name and write it on the menu

e.g.: Kids' Delight for 2 – toppings: 1/2 tomatoes (6), 1/2 pepperoni (6).

12-topping pizza for **4** (*see figure 7*)
- child takes a paper pizza base and divides it into 4 quarters
- child chooses 4 types of toppings and works out that each slice should have **3** toppings each
- ask them to count the toppings on each slice, **establish that ¼ of 12 = 3**
- give this creation a name and write it on the menu

e.g.: Feast Lovers for 4- toppings: ¼ sardines (3), ¼ olives (3), ¼ tomatoes (3), ¼ pepperoni (3).

Y2 Create a pizza menu:

12-topping pizza for **4**, toppings ¾ pineapples and ¼ sardines (*see figure 8*)
- explain that the bottom number of the fraction (4) stands for the number of slices in your pizza
- explain that the top number (3 or 1) is how many slices out of 4 they need to put those toppings on
- child finds the toppings and works out what each side should have
- **establish that ¾ of 12 is 9 and ¼ of 12 is 3.**

figure 6   *picture 7*   *picture 8*

15-topping pizza for **3**, toppings 2/3 olives, 1/3 mushrooms *(see figure 9)*

- explain that the bottom number of the fraction (3) stands for the number of slices in your pizza
- explain that the top number (2 or 1) is how many slices out of 3 they need to put those toppings on
- child finds the toppings and works out what each side should have
- **establish that 2/3 of 15 is 10 and 1/3 of 15 is 5.**

figure 9

Activity

Children are now ready to play pizzeria. Using their menu, they should take orders, create their pizzas and deliver them to their customers. *Buon appetito!*

Follow up

Try creating more menus with a different number of toppings.

Money Minters

Outcomes:

- Y1 counting and recognising coins
- Y2 finding different combinations of coins that equal the same amounts of money

You need: printout of coins (search online) or paper to create your own and scissors

Set up

Look at the coins you have at home and talk about things money can and cannot buy.

Focus

Y1 Cut out the example coins and put them all in the middle. Then:

- pair back and front
- name the coins
- notice differences and similarities
- put them in order of preference and explain why.

When ready, stick back and front together, mix up the coins and put them all in the middle.
Then:

- group them and name them again
- put them in order of value.

50p = 20p 20p 5p 5p

Y2 See above.
Then:

- ask your child to find a 50p coin
- find different combinations of coins that equal 50p e.g.: 50p = 20p+20p+5p+5p.

When ready:

- give your child 50p using a <u>variety of coins</u>
- tell your child they can buy a sweet for 35p
- child gives you 35p and counts what is left (answer:15p)
- now give your child a <u>50p coin</u> to buy a sweet for 35p
- ask how much change you should give back, if needed recap example above
- now ask your child how many pence there are in a pound (answer: 100 pence)
- find different combinations of coins that equal 1 pound.

Activity

Guess my Coin
Choose a coin and allow your child to ask 3 questions, then they need to name the coin.

Y1 *Battle of Coins*
You each take one coin without the other one seeing it, then both hold out the coin in a closed fist. Count 1-2-3, then reveal your coins. The one who has the bigger value, wins the coin. Keep playing with a variety of coins until one player wins all the coins – or until fun.

Y2 *Pound Stretcher*
Play imaginary shops. Whatever your child wants to buy, they can. Draw a picture of these items when purchased, the funnier the better. Child receives 2 pounds, and wants to spend it as quick as they can. Luckily, your shop happens to be the cheapest in town, so all the prices are very low! Whatever your child wants to buy, name the price and give change back. Keep playing until money runs out. Swap roles.

Come back to the real coins you started the session with. See if your child can confidently identify them and count them.

Patisserie chefs

Outcomes:

- Y1 use measuring tools such as weighing scales and containers
- Y2 choose and use appropriate standard units to estimate weight (g/kg)

You need: a mechanical scale, 250 grams of grated biscuits, 100 grams of icing sugar, 50 grams of butter, 1 tablespoon of cocoa powder, approx. 10ml cherry juice (or similar), 10 grams of coconut flakes (if desired)

Set up

Tell your child that you are going to make "no-bake chocolate balls."
Ask them to guess the ingredients (see ingredients above).

Focus

Y1 Children are not expected to be able to read the scale accurately at this level, but you can help by:

- pointing out numbers written on the scale
- explaining that the further the arrow goes on the scale, the more the item weighs
- measuring random items in the kitchen together
- asking if an item will be heavier or lighter than the previous item.

Y2 Help your child read a scale but keep in mind that scale features vary. You might need to adjust the following steps to match the dial on the scale you are using. To begin:

- ask what the first number is after 0 (answer: 50)
- explain that the scale reads in 50s
- read the numbers around the dial.

Copy the number line below then:

| 0 | | | | | 50 |

- count the number of **gaps** between 0 and 50 on the number line (answer 5)
- divide 50 by 5 (answer: 10)
- write the numbers (10, 20, 30, 40, 50) on your number line.

Activity

Before you start:
- measure the ingredients and pour them into little paper bags (see ingredients above)
- put them in order of weight from lightest to heaviest.

Time to get messy! To begin:
- mix the sugar, cocoa powder, biscuits and butter in the mixing bowl
- add just enough cherry juice to make your dough moist so you can mould it
- roll the mixture into balls
- coat the balls with coconut flakes (if desired).

Follow up

Create present packages for family and friends, labelling with the correct weight.

Athletes

Outcomes:
- Y1 comparing lengths for example, long/short, longer/shorter
- Y2 choose and use appropriate standard units to estimate and measure length (cm/m)

You need: paper, colouring pencils, a dice and a tape measure

Set up

Talk to your child about athletics and the different versions of jumps e.g.: high, long and triple. Then ask about animals that jump:
- write a list e.g.: a flea, a grasshopper, a frog, a rabbit, a hare, a jumping spider, etc.
- tell your child that they will be testing their own jumping skills.

Create your record chart:
- divide your paper into 3 columns
- title them: animals, moves and distance
- write down all the animals in the first column
- roll a dice to decide the number of moves the animals are to make
- write these numbers in the second column.

Animals:	Moves:	Distance:
grasshopper	5	354 cm
frog	2	255 cm
rabbit	5	298 cm
jumping spider	4	174 cm
gazelle	6	112 cm
kangaroo	4	510 cm
flea	5	68 cm

Focus

Take a look at your tape measure:
- notice the different measurements e.g.: inches and centimetres
- explain that today we are using centimetres (cm).

Activity

Let the competition begin:
- mark a start line and ask your child to take their position
- call out the first animal from the list, and the number of moves it makes
- child imitates the animal and takes the number of moves
- mark the finish line and measure the distance in cm
- record this distance in the third column using cm
- carry on until the last animal.

Follow up

Create Top Trumps:

Y1 centimetres (*see figure 1*)

Y2 centimetres and metres (*see figure 2*)

- look at your 3-digit distances in cm
- shift your pencil from the end of the third digit going 2 places to the left
- identify that the first digit is the m and the last 2 digits are the cm

N.B.: 2 digit numbers mean the distance is less than a metre

- fill out your Top Trumps cards accordingly.

Play Top Trumps.

Shuffle and equally deal all the cards face down, then:

- both players turn up a card
- the one with the longest distance takes the pair
- continue playing until the winner has all the cards
- reshuffle and play again!

Name:	kangaroo
Distance:	510 cm

figure 1

Name:	kangaroo
Distance:	510 cm
5 m	10 cm

figure 2

Weather Presenter

Outcomes: use appropriate units to estimate and measure temperature
You need: paper, scissors, colouring pencils or pens

Set up

Ask your child to name the 4 seasons. Then:
- ask what the weather is like in each e.g.: cold, hot, warm, snowy, rainy
- create 5 "weather cards" by drawing and writing these words on a card each.

Focus

Ask your child how we can check the temperature outside. Then:
- draw a thermometer like in *figure 2*
- tell your child we use these to measure the temperature in degrees Celsius (°C).

To start measuring:
- look at *figure 1* and ask what the first number is after 0 on the thermometer (answer: 10)
- ask how many gaps there are between 0 and 10 (answer: 5)
- calculate each gap (divide 10 by 5= 2)
- explain it means the thermometer counts in 2s and together count in 2s up to 20
- cut out 5 long cards and draw a thermometer on each (*see figure 2*)
- looking at the weather cards together with your child decide on the temperature for each
- write this as a number on the card e.g.: 41 °C (*see figure 3*)
- using those numbers, complete the thermometer cards accordingly
- mix up the thermometer cards and put them from coldest to warmest (*see figure 4*).

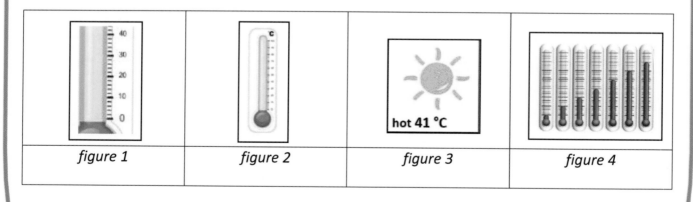

| figure 1 | figure 2 | figure 3 | figure 4 |

Activity

Play Memory with the thermometer and weather cards. To begin:
- shuffle all 10 cards (5 thermometer and 5 weather)
- place them face down on your play mat
- on your turn, face up 2 cards
- if the thermometer matches the weather, you keep them
- if they do not, turn them back face down
- keep playing until all the cards have been paired.

Follow up

Tell your child it is now time for them to present the weather! On an A3 (or even bigger) paper:
- draw the outline of the country of your choice
- stick the weather and temperature cards on and around the map.

Now for the weather!

Mocktail Mixers

Outcome: choose and use appropriate standard units to measure liquid in millilitres

You need: a measuring jug, 100ml pineapple juice, 200ml orange juice, 300ml sparkling water, ice cubes, fruit to decorate, paper and colouring pencils

Set up

Tell your child that today they get to be mocktail mixers. Ask them about their favourite fruit juices, then:

- explain that while fruit is very healthy, they should not drink too much fruit juice as it is high in sugar
- ask them how they could make the mocktail lighter (answer: add water or sparkling water).

Focus

Ask your child to have a look at your measuring jug and point out what measurement we use for liquids (answer: millilitres and litres).

To start measuring:

- look at *figure 1* and ask what the first number on the jug is (answer: 200ml)
- explain it means the jug counts in 200s
- together count in 200s up to 1 litre
- ask how many millilitres there are in a litre (answer: 1000ml)
- ask your child how many gaps there are between the bottom of the jug and 200ml (answer: 2)
- divide 200 by 2 (answer:100) and explain that each line represents a hundred millilitres
- now draw 3 jugs (*see jug 1, 2, 3*) and ask your child to colour them in according to the recipe

figure 1

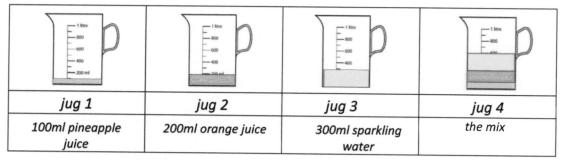

jug 1	jug 2	jug 3	jug 4
100ml pineapple juice	200ml orange juice	300ml sparkling water	the mix

- finally, draw *jug 4* and colour in all the ingredients, reading measurements at each stop.

Activity

Create an icy mocktail using the recipe above. Start with:

- pineapple juice (fun fact: liquids with a higher sugar content will settle down first as they are denser)
- add orange juice very slowly and see if you can create layers
- add sparkling water
- add ice cubes little by little and read measurements at each stop
- ask your child what will happen with the ice and why (answer: it will melt because of the temperature)
- decorate your mocktail with orange slices or any other preferred fruit.

Follow up

Children are now ready to offer their drinks around. Why not try your own mocktail mix for next time?

Cheers!

Geometricians

Outcomes:

- Y1 recognise and name common 2-D shapes
- Y2 identify and describe the properties of 3-D shapes, including the number of edges, vertices and faces

You need: paper, glue, scissors, colouring pencils or pens, some string or yarn, a tree branch and printouts (for Y2 only)

Set up

Look at everyday objects around the room and ask your child which shapes they resemble.

Focus

Y1 Draw a triangle, circle and square then say:

I have 4 sides and 4 corners.
I have 3 equal sides.
I don't have any sides or corners.

Ask your child to guess the shapes (answers: a square, a triangle, a circle).

Y2 Put a ball, a cube, and a pyramid on the table then say:

I have 6 faces, 12 edges, and 8 vertices.
I have 5 faces, 8 edges and 5 vertices.
I have no vertices and am curved all around.

Ask your child to guess the shapes (answers: a cube, a square-based pyramid, a sphere).

Activity

Y1 Create a 2-D chime with the shapes you want to work on (suggested shapes: rectangle, square, circle, triangle). First:

- draw and cut out the shapes of your choice
- while cutting and colouring, ask these questions:

    What is this shape called?
    How many sides does it have?
    Are the sides the same size?
    How many corners does it have?

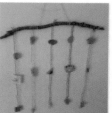

figure 1

Activity

Once you have enough shapes to work with it is time to make your 2D chime:

- take your yarn and cut as many pieces as you want for your stick (*see figure 1*)
- tie these pieces of yarn to your stick
- now take your shapes and stick them on your pieces of yarn
- be creative in designing your chime!

Y2 Create a 3-D chime with the shapes you want to work on (suggested shapes: **pyramid, cuboid, triangular prism, cone, cube, square and triangle-based pyramids**). First:

- draw, or search online for free pintables, and cut out the shapes of your choice
- while cutting and colouring, ask these questions:

  What is this shape called?

  How many faces does it have?

  How many vertices does it have?

  How many edges does it have?

figure 2

Once you have enough shapes to work with, colour them in a different colour for each face.

Now it is time to make your 3D chime:

- take your yarn and cut as many pieces as you want for your stick (*see figure 2*)
- tie these pieces of yarn to your stick
- now take your shapes and stick them on your pieces of yarn
- be creative in designing your chime!

Follow up

Together with your child:

- create a table like below, adding as many rows as shapes you have
- fill out the boxes with the properties of the shapes on your chime.

Name of Shapes* **	Number of Sides *	Number of Vertices (Corners)* **	Number of Edges**	Number of Faces**
*Y1 **Y2				

Car Surveyors

Outcome: understand and create tally charts and simple tables
You need: a ruler, pencil and some paper

Set up

Tell your child that today they get to be surveyors and find out what car colours are the most/least popular on their road.

Focus

Explain to your child that sometimes we use tallies to count. Ask your child to:
- have a look at the tallies (*see figure 1*)
- write down the number each tally represents
- notice what is different about the fifth tally
- think about why this is useful
(answer: makes counting in 5s easier)
- write tally marks for 6, 7, 8, 9 and 10.

For surveys we use tally charts. To make one:
- use a ruler and draw a neat table with 3 columns
- add as many rows as colours you want to work with
- create a simple table (*see figure 2*).

figure 1

Colours	Tally	Total
black	III	3
silver		
pink		
etc.		

figure 2

Activity

Let the surveying begin!
With your tally chart in hand, take your child for a walk up the road or sit by the window and:
- start tallying the cars as you spot them
- make sure you record this information in your tally chart
- find out what colours are the most/least popular by counting the total.

Follow up

Now your child knows how to create a tally chart. They also know how to read information from the table they just created. Ask them to report back on the most/least popular car colours on their road to a member of the family.

You can create new charts for:
• flowers, trees and bushes
• dogs, cats and even people.

Unit 3

Green space: numerals and words
You need: number cards*, 10 small sticks and a larger stick, pinecone or alternative

Tell your child that they need to get ready to play golf. For this, they need to warm up their bodies. Ask your child to do star jumps and give instructions to go slower and faster. Next, ask them to hold their arms out to the side and make circular movements. Movements can start smaller and grow bigger.

Now it's time to set up the golf course. With your child, lay out your A5 cards and read as you thread each onto a stick. Position these "marks" randomly around your course with space in between. Help your child find the first mark and ask them to stand about 1 metre away.

Ask your child to get ready with their larger stick as their golf club and their pinecone as their ball. When in position, they need to hit their ball as close to the first mark as possible. Once happy, they can move on to the next mark.

*Before you head out, write the numbers 1-10 as written words on separate cards. Do this with your child if you can.

Unit 1: Bring me!

Office: everyday items

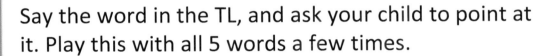

Starter

Show 5 items you would like to teach your child e.g.: a pen, a pencil, a book, a bag and a newspaper. Hold these up one by one, then name them in the TL. Ask your child to repeat the words after you.

Words

Say the word in the TL, and ask your child to point at it. Play this with all 5 words a few times.

Phrases

Put the 5 items away from you around the room. Tell your child how you say "Bring me!" in the TL and then say:

> *"Bring me the book!"*
> *"Bring me the pencil!"*
> *"Bring me the pen!"*
> *"Bring me the bag!"*
> *"Bring me the newspaper!"*

Play a few times then swap roles.

Going further

Once confident you can add *"thank you"* in the TL and encourage *"you're welcome"* as a response.

Available now

YOU CAN TEACH
ME MATHS!

KS1 MATHS TEACHING ACTIVITIES THROUGH STEP BY STEP GUIDANCE
IN ACCORDANCE WITH THE NATIONAL CURRICULUM FRAMEWORK

The Way You Learn Ltd

Coming Soon

You Can Teach Me! A series that helps you teach your child through activities and games.

Just Imagine! A series that uses a storyline to lead your child through various activities.

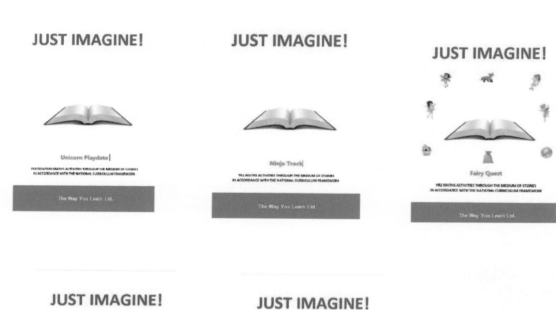

JUST IMAGINE!

Unicorn Playdate

FOUNDATION MATHS ACTIVITIES THROUGH THE MEDIUM OF STORIES
IN ACCORDANCE WITH THE NATIONAL CURRICULUM FRAMEWORK

The Way You Learn Ltd.

JUST IMAGINE!

Ninja Track

YR1 MATHS ACTIVITIES THROUGH THE MEDIUM OF STORIES
IN ACCORDANCE WITH THE NATIONAL CURRICULUM FRAMEWORK

The Way You Learn Ltd.

JUST IMAGINE!

Fairy Quest

YR2 MATHS ACTIVITIES THROUGH THE MEDIUM OF STORIES
IN ACCORDANCE WITH THE NATIONAL CURRICULUM FRAMEWORK

The Way You Learn Ltd.

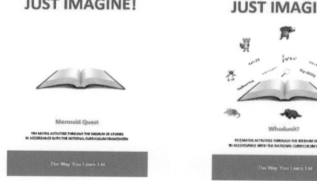

JUST IMAGINE!

Mermaid Quest

YR3 MATHS ACTIVITIES THROUGH THE MEDIUM OF STORIES
IN ACCORDANCE WITH THE NATIONAL CURRICULUM FRAMEWORK

The Way You Learn Ltd.

JUST IMAGINE!

Whodunit?

YR4 MATHS ACTIVITIES THROUGH THE MEDIUM OF STORIES
IN ACCORDANCE WITH THE NATIONAL CURRICULUM FRAMEWORK

The Way You Learn Ltd

The Way You Learn Ltd

As our children entered primary school, we felt the need to support their mainstream learning. We started to design activities for our children that worked well for their needs. In 2019 we formed The Way You Learn Ltd. with a simple goal in mind: to help parents like us support their children outside of school.

We hope you have enjoyed working through this book with your child. We are looking forward to reading your positive reviews. If you have any questions or comments, please email us at info@thewayyoulearn.co.uk or visit www.thewayyoulearn.co.uk

A portion of our profits will be used to reach out to children around the world who may not have access to education. Thank you for helping us achieve this.

Printed in Great Britain
by Amazon